ShowStopper!

By Shonda Carter

Copyright © 2016 by Shonda Carter
All rights reserved. This book or any portion thereof may not be reproduced or used in any manner whatsoever without the express written permission of the publisher and author except for the use of brief quotations in a book review.
Printed in the United States of America
First Printing, 2016
ISBN 978-1539050735
ShoStopping Publishing
A division of ShoStopper Productions, LLC
2810 NE Euclid
Lawton, Ok 73507
www.ShondaCarter.com

Dedication

This book is dedicated to my wonderful husband, Kenneth Carter to whom I submit--but with an attitude...an attitude of praise, an attitude of joy, and an attitude of thanksgiving.

Table of Contents

Foreword.. i
Acknowledgments iii
Introduction: A Different perspective........................…..iv

Part 1: THE W's   1

Chapter 1: Who Needs Video? 2
Chapter 2: What kind of video do I need?....................... 8
Chapter 3: What should your video include and when to use it?... 14
Chapter 4: Why Do I Need Video?................................. 17

PART 2: STOP BEING A SECRET   27

Chapter 5: Survival is the New Success 28
Chapter 6: The REEL You.....….................................….33
Chapter 7: Leave the excuses, take the cannoli!............. 36
Chapter 8: Haters Gonna Hate!.. 38
Chapter 9: Don't Shake it Off, Shake it up! 43

PART 3: Drop Your Message on Millions   53

Chapter 10: What Do I Say?.. 54
Chapter 11: Sharing is Caring.. 62
Chapter 12: Right place…Wrong time? 67
Chapter 13: Going Viral: Chasing the White Whale 73
Chapter 14: Get Rid of the Box 76

Part 4: REEL IN THE DOUGH   80

Chapter 15: Can't Afford Not To 81
Chapter 16: Don't Get Left in the Digital Dust 85
Chapter 17: Effective Video Marketing Strategy 92

Appendix 1 So What Happens Now? 99

Appendix 2 The Videographer's Legal Checklist 101

Authors Biography 106

Foreword

From the moment Shonda Carter stepped out from behind the camera and into the spotlight, she developed her talents into a business and a brand that is known the world over. Shonda spent over 15 years behind the camera learning the techniques that capture an audience's attention and make a client's message resonate. She is the go-to person to make your ultimate video and now she shares her talent and techniques to help you do the same.

In her book, "Showstopper," Shonda's entertaining and quirky style emphasizes the importance of video as your marketing message of choice. "Showstopper" helps you to come out of hiding, eliminate excuses, and showcase your product, services, and mission to the world—on video.

Video has quickly become a means of cutting through some of the distractions that vie for an audience's attention. About 3.5 billion people in the world have access to the internet. Capturing a clients' attention is step one in the process of being heard, acknowledged, and given the opportunity to do business together.

Capturing your ideal client's attention today is not an easy feat. If you don't grab an audience's attention, you won't run a successful business; you will end up with an expensive hobby. Too many people disregard, discount, or simply dump their intellectual property down the drain because they don't take the simple step of capturing it in a tangible format. Video is step one to capturing and legally protecting your ideas, words, and message. All of these are intellectual property. The book's bonus appendix provides a checklist to protect both your intellectual property and legal rights.

Follow Shonda Carter's method to make your message a showstopper and watch your business grow and the money flow.

-Frederick D. Jones, J. D.

Acknowledgments

To my Pastor Paul Craig and all my parents (both spiritual and biological)--Brenda, Vernard, Monica and Lester: I would not be where I am today without the revelation knowledge, monumental support and constant *nose wiping* you have provided over the years.

And a very special thank you to Jade Simmons and Dr. Fred Jones, whose hard work, prayers, and prophetic vision helped remove the scales from my eyes concerning my destiny. www.DrFredJones.com www.JadeSimmons.com

A Different Perspective

My starting statement could be construed as controversial. This may seem farfetched but everyone is a marketer. You all know that when you have something to sell, you will gladly embrace your status and begin to learn how to market effectively. But I believe that opportunities are lost and success is missed when the concept of marketing is misunderstood. One of the biggest mistakes that you can make is letting the word determine your status as a marketer. Whether you are trying to convince ideal clients to sign up for your program, the general public to join your movement, donors to contribute to your cause, and family to embrace a new set of rules, you are indeed a marketer.

But just in case you are still not buying it, I believe this example of the greatest marketer of all time (even though he is the last person you would think of) will clarify my point. So who could it be, no it is not Muhammad Ali, Steve Jobs, Walt Disney, or Mary Kay Ash—remember--I said he would be the last person you would think of. His name is Jesus Christ and without a doubt he is the greatest marketing guru ever. To some people, the thought of Jesus engaging in worldly endeavors such as marketing seems almost ludicrous. But, didn't he flip the tables when people were trying to sell in the temple (Matthew 21:12)? This is why he is the perfect example of what the term marketing really means:

Marketer—A person whose duties include the identification of goods and services desired by a set of consumers, as well as the marketing of those goods and services on behalf of a company (Businessdictionary.com).

Jesus is a marketer because he recognized that *the world* needed *salvation* and desired *abundant life* so He offered his goods and services on behalf of the *Kingdom*. Viola! This definition is only an attempt to look at marketing from an unexpected perspective, it is hardly meant to minimize the ultimate sacrifice that was made on behalf of mankind. However, if you take the time to study the Bible from a marketing perspective, you cannot help but be impressed with his skills in the execution of simple steps revealed in Marketing 101.

Jesus focused on a specific group of people, or as we like to say in marketing: found his niche (Matthew 15:24). He positioned himself as an expert and communicated his core message with a direct, laser-like focus that grabbed people's attention (Matthew 4:17). He never failed to deliver value to his followers (John 21:25) and he always required a response, aka a call to action.

If you move past marketing basics, you will see another area of his expertise: relationships.

Relationship Marketing is a form of marketing that emphasizes customer retention and satisfaction rather than shorter-term goals like individual sales (Wikipedia).

As a master at developing relationships, Jesus started with 12 and currently has 2.5 billion (1 out of every 3 people) over 2000 years later claiming to be one of his followers. Once you begin to look at the concept of marketing from a different perspective, you will easily recognize and overcome the common types of challenges that every business, ministry, or social movement must endure along the journey to success. For example, despite how badly the world *needed* his message, miracles and services, even Jesus had to constantly deal with the pain of rejection.

Today it has become popular to incorporate spiritual principles into business practices, therefore, whether you are an entrepreneur, a leader, or a legend, you will not be exempt from developing and using marketing skills.

Injecting story and personality into a company's online presence sets your message apart from the others. When people know, like, and trust you, a relationship develops that is crucial for building loyalty and spreading your message. Because today's society is visual, nothing is better, faster, or more effective than video to *reel* in the results you need for success.

Part 1:
THE W's

1

Who Needs Video?

"No matter what you do, your job is to tell your story."

—Gary Vaynerchuck

Let's start with the obvious; do you want or need money to keep the doors open to your business? Oh what the heck, let's just be controversial--do you want to earn a truckload of money by doing what you love? If the answer is yes, then you need to make video. Do you have a message or mission that is so important that it could change the world? Then you need video. Are you an expert in your industry but only your friends and family know how great you are? Then you need video. These questions are not meant to be insulting but it never ceases to amaze me how hard I have to work to convince some people of the importance of video. My understanding happened when a business coach told me not to use the word *video* because the word is scary. I was confused by this statement until I considered that I could effectively market what I do. Because there is an actual phobia (scopophobia) associated with my best friend, the video camera, I realized why people are afraid of a video camera. I also realized that the word also conjures up images of work and no one wants to invest in anything associated with phobias and work.

So let's start over, you don't need video. No, no, no, you want a video because it saves you time. You do not have time to introduce yourself to thousands of potential clients every single day. You also need to tell your story. At numerous speaking events, the first question asked is always: "How did you get into your business?" Or "What made you get started?" People want to know your story. You need video, therefore, if you are an author, speaker, entrepreneur, social innovator, non-profit organization, event planner, or in ministry.

Authors

Even if your book is still in the planning stages, it is best to start doing your videos now. Both you and your title have to be located on the internet and videos are more likely to put you on Google's first page (they own Youtube). The more videos that you have on your book subject, the quicker you will establish yourself as an expert and build the human connections that are crucial for book sales.

I made "Showstopper" available for pre-order and it reached the status of a #1 Amazon best seller in a matter of hours because my videos created my fan base and established me as expert in video. I am respected because I am crazy, quirky, cool, and trusted.

Video will not get you instant trust but it will grant you the power of instant connection. When people see your face, body language, and hear your voice they make an instant decision on whether they like you or not, which effects their decision to buy or not. The decision to do video will make your audience notice you; you will stand out in an industry of introverts.

Speakers

Being a great speaker today is no longer good enough; you must not only showcase your talent but you must also standout from those that are just as talented as you--the purpose of the sizzle reel. The sizzle reel is a 3-5 minute video that combines visuals, audio, and messaging to create a fast paced, stylized overview of a product, service initiative, or brand. You should invest in this type of video if you want to get booked and make money. I use the word *invest* because this type of video should be done by a professional.

Don't ignore this powerful tool because you do not consider yourself to be a professional speaker; as an entrepreneur, you may already be aware that not many events are eager to pay for a keynote speaker; most speakers prefer to now sell from the stage. Getting booked on a stage in front of hundreds of potential clients is the fastest and most effective way to market yourself and your services. You can also use video on stage to keep your audience engaged, to display and explain your points, and entertain.

Entrepreneurs

When video shares your message to millions, it not only creates brand recognition and explains your products or services, it helps you make money.

Social Innovators

Healing the world is not an easy task. And neither is explaining your new concept or cause to every passerby. Video creates consumable bite-size stories that capture the heart and display the humanity of complex social causes. Powerful video has been hailed as the spark that starts a fire or keeps it going. Video forms a visual memory or imprint that each view refreshes and strengthens. News coverage consistently relies on images to hold an audience's interest and to sometimes influence how events are depicted and remembered.

Non-profit Organizations

Even though you are not in business to make money, you still need it to effectively help in the area to which you are called. Creating a video can evoke the emotions necessary to raise money with a strong call the action.

Event Planners

Your mission is to fill seats. Sounds like the same mission the movie industry faces. Hmmm, I wonder how they would do that without the creation of a movie trailer. Your video can fill those seats too.

Ministry

Jesus was a master storyteller but he was not the only one. He knew that the power of combining a story with visual action would give his message a greater impact. If your ministry is old school--sermons and music only--your message could be reaching less than half your audience because 65% of the population are visual learners. People need learning techniques that accommodate their style. I left so many churches and gave up on the whole Christian lifestyle because I was frustrated with the message's delivery. I am happy that in 2004, I was truly saved when the film "The Passion of the Christ" opened in theatres. I was defenseless against the multisensory love story that awakened my senses dulled by years of the *listen only* approach. Story, visuals, and powerful music are very effective when it comes to punctuating and expounding the truth.

If I haven't listed your profession but you still want to make money, reach a big audience, and make a big impact, you need video. No matter the job or profession, there is no industry that video cannot enhance.

2

What kind of video do I need?

"Engage rather than sell... work as a co-creator, not a marketer."-Tom H.C. Anderson

It might not be easy to get discovered in Hollywood but it seems that writing a movie is beyond simple. That is because films are built on the same three basic components: character, desire, and conflict. The formula may be straightforward but the genres of movies available today are complex. Online videos are increasing as technology evolves; for example, The Talking Head, interview style, live streaming, Tip Series, Photo Montage, Sales Video, Testimonials, Explainer Video, Teaching Video, Animated, and Sizzle/Speaker Reels. These types of videos are very effective but if your goal is to establish yourself as an expert in your field, there are three essential video types that you must have: Showoff, Showcase, and ShowUp.

SHOWOFF

"Knowing a person is like music, what attracts us to them is their melody, and as we get to know who they are, we learn their lyrics." —Unknown

These types of videos require you to think about a relationship--you have to showoff who you are. Imitation may be a great form of flattery but when it comes to these types of videos, trying to be someone you are not is one of the worst things you can do. Do you remember the show, *What Not to Wear (*a reality television makeover show*)*? One of the episodes featured an analogy I will never forget. Clinton Kelly, one of the co-hosts, explained why a 55-year-old woman should not be shopping in the pre-teen section even though she could still fit into the clothes. His explanation had nothing to do with ageism but everything to do with perception:

> "When a guy sees a woman from behind and she is dressed like a 20 year old, when he approaches and she turns around he will be disappointed that she is 50." –Clinton Kelly

No one wants to feel like they have been lured into a relationship by means of a fictional online persona--not in love or business. Remember, your audience has to know you before they can make the decision to like you and ultimately trust you. These videos showoff your uniqueness and they can save you vital time and energy by attracting the right people and repelling those who could become a nightmare later.

SHOWCASE

"After years of hard work & practice I am now an expert in sarcasm. GO ME!"—Garfield the cat

This is the video that will showcase your authority and expertise. It is usually embedded on your sales page or website; think of it as a job interview. Your viewer has already checked out your stats, now the viewer wants to meet you in person. Please do not let the word *interview* convince you that this is the time to be stiff and stodgy, quite the opposite. Now is the time to showcase your uniqueness and expertise. Do not disappoint. The last thing you want is for fans to fall in love with you on social media and feel like they are seeing a totally different person when they finally visit your website. Show the viewer the results he or she can expect if they work with you.

We live in a very judgmental society. I know that most people don't want to believe that but it is the truth. Most don't want to admit this truth because we have been made to believe that all judgment is wrong. I breathe a sigh of relief when I visit a dentist and the dentist looks the part: white coat, great smile, and confidence, old enough to be a dentist. I don't care a great personality—when I am in pain, I want the pain to end and I want it to end immediately. I need reassurance that you are the one who can get the job done.

Let's look at this with business eyes. If your brand, title, or tagline has the word *million dollars* in it, I expect to see a professional, high quality video featuring someone who is confident that she or he can help me make a million dollars. You have to walk, talk, look, and exude the part. This is not necessarily exclusive to millionaire brands. Any time you ask people for money, whether it is for services or to help fund a non-profit, you must remember: the higher the quality, the higher the trust factor must be. Once you are trusted, your prices no longer become a potential hindrance.

ShowUp

"You can pretend to care but you can't pretend to show up."—George L. Bell

Is there a party over here? Well ok, it is not always a party. More than likely it will be a live event, a webinar, or a fundraiser. The desire is the same; you want an audience. This is when the magic of a promo video happens. Promotional video does more than promote an event; it can also promote you and your genius. A speaker reel, for example, can feature how fantastic you are and give people a sense of what they might miss if they don't attend your event.

Don't mistake a promo video for an opportunity to make a commercial. People are very savvy; no one sits for commercials. I was once called out during a Question & Answer segment for making that statement. I defended my stance because some people only watch the Super Bowl because of the commercials. I quickly explained that due to pure marketing genius, these are no longer commercials; they are now an experience. Yes, they are exciting, entertaining, and fun but the brilliance of the campaign shows up when you feel a sense of loss because you didn't show up to watch. Think of all the conversations you won't participate in after the game is over. Maybe you can't afford to run your ad to reach an audience size of the Super Bowl but that is no excuse not to put the same effort into your marketing.

Professional doesn't equate to boring and dry. Although these videos are used for very specific outcomes, they can still be fun and creative. If your video is dry and boring, people will not watch it long enough to get any value out of your message.

3

What should your video include and when to use it?

"Your vibe attracts your tribe."

Not knowing how to use video marketing is a mistake but not knowing what it should include is an even bigger mistake. Thanks to social media we must say hello and embrace connection. If you want to have success, a personal connection to the camera is a must. No matter how impressive your logo or assembly line of products may be, it will always fall short of grabbing a viewer's attention.

A video has the power to give a product or space a personality, a vibe. The story and emotions with music and quick shots can display the atmosphere you want to capture in under a minute. The basics include music, brevity, humor, a great opening, face time, story, and a call to action.

Music

Music creates the emotions you want associated with your brand and drives the video forward. The right music makes your video feel like it is shorter than it is.

Brevity

I caution my clients about the length of their videos—no more than two minutes. Exceptions exist but professional help should determine when, why, and how.

Humor

No one complains about too much humor. It is a great way to get people's attention, keep them engaged, and increase the possibility of getting your video shared. Attention spans are now short-- grab it from the beginning. Get right to the point.

Face Time

Social media demands connections not commercials. Having your face synonymous with your video marketing is powerful. It creates familiarity and builds trust.

Story

All great videos have a story element. You have to talk to your audience, not at them. When your videos tell a story, you invite your audience to become a part of the experience. Stories sell so tell yours well.

Call to Action

I always thought the viewers would read my mind and just instantly know what to do. It wasn't until someone became angry because I did not tell her what to do next that I realized that I needed a call to action. Never make a video without a call to action even if it is solely to get your viewer to check out more of your videos. Do not make it subtle; make it punchy and make it clear: "Hey, if you love what you are learning , stop everything and give me a quick review on Amazon!"

4

Why Do I Need Video?

"Video is everywhere.

Small businesses who ignore it, 'do it at their

peril."- The Guardian

What do authors, entrepreneurs, speakers, social innovators, musicians, small business owners, instructors, life coaches, restaurant owners, accountants, business coaches, big corporations, dentist, mechanics, scientist, social media celebrities, tae kwon do instructors, etc., have in common? They all need video but only if they want to grow their business, get booked, and get paid.

Whenever I do more research on the effects of video for the future, I find myself humming the song, "Video Killed the Radio Star," by the Buggles. I can't help but think of music artists that once topped the charts that are now nowhere to be found, at least to the general public, due to a change in the interest and taste of the music industry. One of the most significant changes was the rise of the MTV Era. When it arrived it turned the music industry upside down. Suddenly, it no longer mattered how great or talented an artist was. If an artist refused to believe that a music video was essential to their career, they were not going to succeed. I now see the same refusal to believe in video marketing in business, social movement, and even ministry genres.

Anyone who has hired or talked with a business coach is probably familiar with the term *pain point*--a problem, or need, that a business or company aims to solve. This term is one of the catalysts responsible for ditching my comfort zone behind the scenes and stepping into the spotlight to promote my business. I never considered myself to be a saleswoman yet I was usually required to sell fellow entrepreneurs on the benefits of video. This always angered me because then I was practically working for free so my resistance was high. I grew exhausted and bitter. I even gave up for a little while because I did not want to work that hard to save another's business.

It wasn't until I decided to step in front of the camera and start doing videos that explained and entertained the subject of video marketing that I started to see a shift and ease in my efforts. I get to showcase my own unique personality, and no longer feel that I have to imitate a cheap car salesman.

I do realize that there are some of you out there who need more than just a few stories or opinions to educate you on the effectiveness on video, so here are more reasons why video is now a must have.

"Stick to the facts, ma'am"—Joe Friday, Dragnet

According to Cisco.com, by 2017, video will account for 69% of all consumers' internet traffic. Video-on-demand traffic will double by 2019.

Video is no longer the future, video is now. Various studies show more than half of companies are already making use of video marketing–this figure is predicted to rise. Nielsen claims 64% of marketers expect video to dominate their strategies in the near future.

The number of mobile phone video viewers in the US is projected to reach 137 million by 2019. Consumer internet video traffic will be 80% of all consumer internet traffic across the world in 2019. Because of the race for social media supremacy, Facebook is generating 8 billion video views daily and Snapchat is generating 6 billion. Can you imagine what whose figures will look like by 2020?

"I only believe in science."-Esqueleto, Nacho Libre

It doesn't take a scientist to tell you that video makes it possible for people to hear and see you. What will peak some curiosity however is that video also makes it possible for people to *read you*. By using the phrase *read you*, I mean that your audience gets to see in a video what text alone could never show--your body language. UCLA research has shown that only a mere 7% of communication is verbal. What about the other 93%? Glad you asked. Thirty-eight percent comes from your tone of voice and the remainder of your communication comes from body language. That is a whopping 55%, which is why video is essential when you're trying to engage with your audience on a personal level.

Whether we realize it or not, we were designed to read and look for faces. Dictionary.com defines the word *pareidolia* as our imagined perception of a pattern or meaning where it does not actually exist, as in considering the moon to have human features.

Research reveals that our brains are built to establish trust through face-to-face contact and the majority of that trust falls on the eyes. Turns out that scientists have proven the famous quote from William Shakespeare about the eyes being "...the window to your soul." Research discovered that patterns found in the iris can indicate whether we are warm and trusting or neurotic and impulsive. This type of research does much to explain why we need to see a person and know their story is such a vital part of many presentations and television ads.

"Marketing is a contest for people's attention."
--Seth Godin, entrepreneur and author.

Attention is the new currency and because of the growth of social media, it is getting more difficult to gain people's attention. Video is becoming more popular because it gives people the opportunity to absorb a massive amount of information in the most effective way. If, however, you pump out typical ads with nothing more than information, you do it at your peril. Yes, video can be used to educate consumers but when we use it to market in a more memorable way, we reveal its real power.

Because online attention spans have diminished, I always check the length of a video first to determine if I will watch it. If the time clock says anything beyond two minutes, I will not watch it (unless I already know your videos are great). Ok maybe I will, if I am waiting in a long line, but other than that, no way! I should also warn you that most people don't want to spend their precious few minutes on something you think they should learn. I can't say that I blame them; we are bombarded constantly with information. Because of this, people no longer seek to learn; they want to feel. And there is no greater medium than video to hold the attention span of your audience with the help of entertainment and storytelling.

Video is the new website. Consumers watch video more frequently, so they now expect it. From a marketing perspective, video can be used for more uses than just attracting your audience's attention. The following rationales stress the importance of video:

1. **Showcase:** Video showcases your uniqueness and sets you apart from your competition by allowing people to connect with your story.
2. **King of Conversion:** Research shows that 71% of marketers state that video conversion rates outperform their other marketing content.

3. **Sales:** Video does the work of selling for you. Think about how many times you may have walked passed a product before you saw a commercial for it and made the decision that you could no longer live without it. You are not alone, a study conducted by Animoto shows 76% of people are more likely to buy a product or service after watching a video that explains the product.
4. **Search Engine Optimization:** Video helps your website get discovered more easily by Google and other search engines. Did you realize that Youtube is the 2nd largest search engine? Youtube processes more than 3 billion searches a month and 100 hours of video are uploaded every minute. According to Pinnacle Marketing, it is bigger than Bing, Yahoo!, Ask, and AOL combined.
5. **Social Proof:** Video testimonials provide credibility for any business. Testimonials provide a shortcut to the three main factors needed: *know*, *like*, and *trust*. One to the best examples of social proof is the canned laughter used in TV sitcoms that triggers others to laugh at the same things.
6. **Celebrity Effect**: Because of the extensive growth of social networking, a plethora of self-made celebrities have surfaced. This

new breed has become popular by creating videos that get shared. As marketing experts know, fame gives people an instant halo of credibility and trustworthiness, which is once again essential when doing business online.

7. **Different Learning Styles:** Blogs are great but they only cater to one type of learning style. Video is greater because it can incorporate all learning styles. We process visuals 60, 000 times faster than we process text. Whatever the purpose of your video, your basic goal will always be to keep the attention of your targeted audience. Why not ensure a home run with the power of video? Seven learning styles with the mere push of a button-- that is a formula that cannot be beat.

8. **Everybody's doing it.** Video production costs have fallen significantly making the technology available to everyone, which includes your competition. Studies show that there are now more than 50 million small businesses on social media. The changes made by Facebook, including Facebook Live, are responsible for a surge in video views. This surge equals more than 8 billion views per day.

Hopefully, by now you have now joined the 87% of online marketers who according to Outbrain.com agree that video is no longer a luxury but a necessity. But just in case there are still some doubts, let's get more specific about the industries that video can enhance. I have covered the majority of reasons for the necessity of video for your movement, marketing, or messaging, using facts and figures, scientific, psychological, and marketing research. And even though the effects have dissipated since its 90's peak, I even threw in some good old-fashioned peer pressure just for nostalgia's sake. Before you become 100% sold, however, I realize that my information may have generated questions. Sometimes more information brings more questions such as:

- What kind of video do I need?
- What do I say?
- What should my videos be about?
- How much will they cost?

Producing high quality creative content can be a challenge. Technical skill set, equipment expense, and a lack of creativity could present barriers to video marketing success. I will be covering some simple solutions to all your queries. After all, "I'm your huckleberry." –Doc Holiday, *Tombstone*. I will provide the help that you need, so let's get started so I can keep you from being left in the digital dust.

PART 2:
STOP BEING A SECRET

5

Survival is the New Success

"It always seems impossible until it is done."—

Nelson Mandela

"Hmm, it sounds like you are selfish and self-centered to me. Maybe it would help if you got yourself off your mind."

I will never forget those words I heard 15 years ago when I starting seeking help for my battle with stage fright.

After the initial sting subsided I remember the next emotion eagerly waiting in line to rear its ugly head was anger. I was down right miffed. After all, I had traveled across two states to attend this special ministry meeting. I was genuinely seeking help and my motives were pure. How dare you say these things to me, Minister Andrew Wommack. YOU DON'T KNOW MY LIFE!

The message was rough but trust me I needed to hear it that way (rough, raw, and uncut). Because once I stepped out of my emotional turmoil (and finished listening to the rest of the sermon), I realized he was correct.

self·ish

(of a person, action, or motive) lacking consideration for others;
concerned chiefly with one's own personal profit or pleasure.

self-cen·tered

preoccupied with oneself and one's affairs.

At the time, I did not fully understand the gravity of what was about to happen in my life, but I knew that whatever *it* was, *it* would never happen if I kept having mental meltdowns everytime there was an opportunity to reveal some of my hidden gifts, talents, skills, or heaven forbid--expertise.

"According to most studies, people's number one fear is public speaking. Number two is death. Death is number two. Does that sound right? This means to the average person, if you go to a

funeral, you're better off in the casket than doing the eulogy."—Jerry Seinfeld.

I know that Jerry Seinfeld was telling a joke, but I remember doing just about anything to get out of giving a report, posing for a picture, or doing anything that required having human eyeballs focused on me. It wasn't until I heard those fateful words over 15 years ago that I finally realized that if I had any hope of fulfilling my destiny it could no longer be about me.

Some of you may be battling feelings of unworthiness or uncertainty when it comes to your gifts. You may not believe you are ready or good enough. There are some of you who believe that nothing is waiting to reward your steps of bravery but mockery or good old-fashioned failure. And as harsh as it may be, I have to give you the same advice: "You are being selfish and self-centered. Maybe it would help if you got yourself off your mind."

"The purpose of life is to discover your gift.
The work of life is to develop it.
The meaning of life is to give your gift away".—David Viscott

It is time to put things into the proper perspective. You have something that needs to be shared; once you accept that it is not about you, you will find it harder to shy away from stepping into your spotlight because of fear of being misunderstood, or getting it wrong.

I say all of this to let you know that I know video can be scary. Despite getting revelation knowledge smack dab in the middle of a church service, it still took me quite a bit of time to get comfortable with being the face of my brand. When I finally managed to get over my fear of the stage, I ran into another problem:fear of being misunderstood. I still laugh whenever people refuse to believe that in my B.B. (Before Business) life I was not an entertainer. I used to cringe when meeting a person that saw me on stage or video. Combine their state of shock with my being an introvert and you can see a recipe for disaster being stirred right before your eyes. If that wasn't enough, I had also managed to conjure up a whopping dose of insecurity because I did not have the personality of natural born entertainer. I don't know which was more irritating--the people that were disappointed to meet me (of course I am exaggerating; I don't know if that is what they thought. Remember this is all going on in my head). Or perhaps people believed that I never get nervous because I have the ablility to command a stage.

    I write all this to let you know that you are not alone. I have either personally been there or have worked with clients who have been right where you may be. There is no excuse that you can give, or situation issue you may believe that will prevent you from becoming a showstopper and creating phenomenal videos that convert like crazy. Especially since, you already possess something that makes your brand outstanding--distinction. Just because you may not have recognized it yet does not mean its not there. So let's begin the process of finding it along with creating a show stopping video:

#1 ~~Buy the book SHOWSTOPPER~~
Look at that, you are already ahead of the game.

#2 Get over your fear of the death ray--I mean the camera. If you suffer from fear of the camera, this step will not be so easy

But trust me it's worth it. You owe it to yourself, your brand, and the people who are awaiting you.

6

THE REEL YOU

Just be yourself! When I hear an expert behind the camera give that advice, I feel nauseous because that advice is almost impossible to follow unless you are naturally a camera-confident individual. You should always be authentic--there is no greater fake detector than a video camera but *just be yourself* is hardly helpful.

Every relationship starts with an introduction. Can you imagine showing up for a date and discovering your potential match has not showered, hair not combed, teeth not brushed? Video marketing takes the same amount of work required for any successful relationship. When you go out on a first date, most of us present our best selves--am I right? It is not being fake, so the sooner we eliminate that idea, the better time you will have in front of the camera. You may think I am going a wee bit overboard but I have had to have these types of conversations.

Some people have talked themselves out of pursuing video marketing because they believe that it requires them to be a fake. I once had a conversation with a woman who was not comfortable with the idea of video because it would require her to wear makeup. Now, normally I would have taken the time to coach her about this area or given alternatives but before I could stop, I literally laughed out loud. I laughed because this gorgeous woman was on social media with the most stunning photos I had ever seen wearing flawless makeup. After my laughter subsided, I asked her if she felt like a fake every time she got ready to go onstage to talk about her business by putting on a lovely outfit or getting her hair and nails done. She didn't even bother to voice her *no* because she quickly got the connection. So while I despise the idea of trying to be someone else, I caution against *just be yourself* in video marketing. When that camera is in your face, it is time to become your *reel* self; i.e., the best version of yourself. I know this sounds a bit artificial to say but just like on a date, if you don't learn to *turn it on,* when it comes to video, your audience will quickly click you off out of boredom and bounce on over to someone else's video.

Time is precious, so if someone is spending it on your video, they definitely expect you to make it worth their time. Despite that you have explained the benefits of your services for 10 years, you still have to show up full of excitement and enthusiasm. If you are not motivated, how can you expect your audience to be? If someone has clicked on your video that means they have taken the first steps to connect with you. They actually want to get to know you but that won't happen if you are trying to hold on to a boring persona under the guise of trying not to fake it. Just how do you become the *reel* you, then? I am glad you asked.

7

Leave the excuses, take the cannoli!

If you haven't noticed, I speak fluently in movie quotes. The quote that I parodied for this chapter title was originally used in the 1972 epic *The Godfather*. The original quote is "Leave the gun, take the cannoli." This line resulted from improvisation and means that some things are just part of the job. The future is sweet--still available to be enjoyed—therefore, it is more important.

It is time to dump the excuses and realize that striving for success will require you to leave your comfort zone. The first line of defense usually starts with: *But I am an introvert*! I hate to plow through your defensive line--ok, I take the back. I love doing it but being an introvert means that constantly being around people can feel like the life is being drained out of you. It has nothing to do with being shy or unable to speak on camera. I know this because as I previously stated, I am an introvert and so is Meryl Streep, Michael Jordan,

Hillary Clinton, and Barack Obama. While I was thankful to be in such great company, I was not happy that I could no longer hide behind the title to prevent me from doing my job. And neither should the thought of potential criticism.

8

Haters Gonna Hate!

I teach an adult class on video marketing at a local technical school. Most people sign up for generally the same reasons: to learn to become the face of their brand, to learn the technicalities of video, etc. The amount of resistance I encounter when it is time for my students to actually make a video never ceases to amaze me. These same eager students that were hanging on my every word now view me as their mortal enemy; a fight always ensues. Which is why, I always have to start each class with a Mr. Miyagi (*The Karate Kid*) Method of teaching. Before we continue, therefore, let's make a list of some of the scariest things you can think of about being on video. List at least five.

If you are anything like most people (me included), one of the things you listed was a fear of being judged. Aristotle stated, *"Criticism is something you can easily avoid by saying nothing, doing nothing, and being nothing."* If you are committed to creating a business or making the world a better place, you can deal with anything directly. Numerous articles assert that criticism can be beneficial, and I understand what this research is trying to prove but I disagree. One of my favorite bible scriptures asserts that a lack of knowledge destroys people. To resist the temptation of remaining a secret because of fear suggests that you learn the difference between criticism and critique.

Criticism	Critique
• Looks at what you don't have	• Looks at what is working
• Condemns what was not understood	• Request clarification
• Sounds Cruel and sarcastic	• Sounds kind, honest & objective
• Vague	• Always specific
• Focuses on flaws in the person instead of the message	• Concerned with the message portrayed

I often receive comments on my videos whether I like it or not. I have had to learn how to tell who is honestly trying to help me become better and who is trying to belittle me. If you plan to step out or you have been thrust into the spotlight, this skill is essential for your pending success. If you are being lovingly critiqued, don't fall apart, take notes on what you can change and make your next work better. If, however, you know that you are being criticized, don't change a thing. Don't fall into a Catch 22: A requirement that cannot be met until a prerequisite requirement is met, however, the prerequisite cannot be met until the original requirement is met. When people are dead set on being critical there is really nothing you can do to make them happy.

Another snare to avoid is the comparison trap because *there is no comparison*. We are unique and wonderfully made; therefore, any comparison to others will be unfair at best and wounding at worst.

Unfortunately, comparison is another sign of visibility. People really can't help making comparisons and they don't intend to be mean-spirited. They are trying to figure out what you do. Whenever our brain is involved, we usually seek the easy way when it comes to understanding things; other people then try to put you in a box.

When I first started out, I made funny Christian videos but no one could accurately describe exactly what I did. Comments included, "She is the next Tyler Perry." I didn't put up too much of a fuss because even I was at a loss for words. I was having fun so I just let them talk. It wasn't until I became clearer about my messaging and started making videos for my own business that my defensive fangs started to come out. It wasn't so much that they compared me to others; it was the manipulation that came with it:

> "I love your videos but I think that you should tone it down a bit. People might be scared off if they think they have to be as crazy as you to make a video. You should check out some of so-and-so's videos, you two do almost the same thing but he is a bit tamer."

> "Shonda, aren't you worried that you will scare away potential clients with your crazy videos? You should be like so-and-so."

I actually fell for it for a while and made one of those vanilla backgrounds, talking head videos, only to get the following message:

> "Wow Shonda, I just saw your video and I did not see you in it at all! I might be totally out of line but I am giving you a snooze alert!"

I made the mistake of letting other people put me in a box. But I believe it is far worse when we fall into the trap of constantly comparing ourselves to others. Can you imagine the ridiculousness of sending a message like that to Kevin Hart:

> "Hey, I caught your act and yes you are very funny but perhaps you should try being a bit less animated. You should check out Jerry Seinfeld, you both make people laugh only he is just a bit tamer. I really believe it would help you advance your career."

We have each been given gifts and talents that need to be shared with this world. Trying to compare you with others, therefore, is futile. I realize that when you are first starting out you need some time to build up your defenses against the attacks that may come from others; so much of this advice is sometimes easier said than done. The real hard work begins when the attacks are coming from within.

9

Don't Shake it Off, Shake it up!

Before I started in this line of work one thing guaranteed to drive me absolutely bonkers was hearing about a supermodel who complained about her looks. I did not believe she had the right. You could imagine my surprise when I finally noticed the dirty looks that I earned when I dared to complain about a few pounds I had gained since I had reached a certain age. I gave birth to four children and maintained the size I wanted for many years, and now the old bod was starting to do its own thing, so I felt that I had a right to complain. But after receiving the *looks that could kill*, I realized that from the outside looking in, it appeared that I had nothing to complain about. It was that day that I learned that the truth cannot be seen if our crazy perception is blocking our view.

So here is Mr. Miyagi task number two: Make a list of everything that you believe is physically wrong with you, or better yet, how you think you look on video. Don't worry, I will go first. For starters I broke my tooth 24 years ago so I just knew that despite how funny my video was or how many costumes I donned, I knew that everyone was going to finally notice that my front tooth was fake—great now you will probably be looking for it. Ok, let's keep going, I am not getting any younger so I have noticed a few lines and wrinkles. I know that people are probably wondering why this 112-year-old looking woman is creating so many videos. And now--for flaw that has caused me to delete a video faster than a speeding bullet--my neck! If you have seen any of my videos, I know you couldn't help but notice that my neck is the length of a giraffe's. It is so hideous and long I don't understand why the Guinness Book of World Records hasn't contacted me yet. Yes, my list and flaw descriptions are a bit on the dramatic side but if you catch me on a bad day, some of those thoughts still come into play.

So let's get to your list. I want you to take it and read everything you listed to one of your children, your spouse, or a close friend and tell them this is how you really feel about them. Did you do it? Oh come on, don't be a chicken. What's the harm? Ok, maybe you feel funny just saying negative things to them out of the blue but the next time one of your children shows you a school paper I want you to say, "I hate everything about it." "Why did you even bother?" "Do it over, you messed up." The next time your friend or spouse asks "How do I look?" I want you to say, "You look stupid" or "You look like a hideous beast. Oh yeah, and did I ever tell you how much I hate the way you sound." Are you getting the picture? This negative self talk is the reason I created the video golden rule: If you won't say it to anyone else, you are not allowed to say it to yourself--end of discussion.

Time to get down to the basics: Not a single one of us is perfect. Despite my playful way of dealing with things, I do realize that there are real issues that stop some people from getting in front of the camera. I know that we are focusing mainly on business and building a brand with video but it goes much deeper than this. There are some people who won't allow their photo to be taken and they don't realize that their behavior is robbing their friends and family of precious memories. And it is all because they refuse to accept or love the not so fantastic parts of themselves. Not going to lie, this is going to be a fight but everything worth having or doing usually is. So here is where to start:

Choose Your Battles Wisely: If there is something you can do about it—do it.

I am 44 years old and I still have acne. It's not fair, it's not right, and it is not ok. Do I wish I had skin that didn't flare up every time I eat too much sugar? Absolutely! But I discovered makeup concealer and with a couple of swipes of the sponge--Boom!--instant camera cure. I know this example is a bit simple but sometimes we just make things too difficult. I once used a two-week juice cleanse to detox and clear up my skin whenever I needed to make a video. The result usually ended with the video not getting made, leaving me feeling worse than I did before I detoxed. The only result we get when we allow our flaws to stop us is feeling even worse. The flaw is still there but now we beat ourselves up because our task was left undone. This brings me to strategy number two:

If you are not going to do anything about it, stop bringing it up.

One of my favorite quotes is from an exercise DVD I bought years ago: "You're never going to change until you want it more than you want to stay the same." It's time to make a decision. If you say you want to lose weight and you are going to do what it takes to complete your goal that is great. But if you are not, it's time to realize you have been using this unreachable goal as an excuse. Let it go, learn to love your size, and rock your videos.

If you can't fix it, all you can do is stop fixating.

I deal with the perception of a hideous disfiguration known as my long neck on a semi-regular basis. There is nothing I can do about a long neck. Is there such a thing as shorten-your-neck surgery? Just kidding, I wouldn't pay for it or consent to it if there was. Especially since I know that the brief fixation with my neck is all in my head. I never even thought that way until a close friend said my neck looked like a giraffe's when I was caught in an awkward pose for a picture. After that I couldn't shake that voice telling me to beware of my neck every time a camera was in range. Fortunately, I did not let that little voice stop me. I did the opposite. I made a video making fun of myself. Now when that voice tries to come back with those old tricks, or even when I see a giraffe, I can't help but giggle. I know that everyone has his or her own self perceptions to overcome about being on video but you have to learn to embrace those perceptions or change them. There is no other option. Notice I did not mention *complain about* and *avoid*.

At the threat of being once again considered overly sarcastic, I am going to have to drop some groundbreaking knowledge right now. People really don't pay close enough attention to keep track of all your flaws, well, unless you have haters but you are not supposed to be paying attention to them anyway. You are the one who has to set the focus, and you would be wise to set the focus on your message instead of your insecurities.

Set Your Focus

If you want to make a great video, it is essential to take the zoom lens off your flaws and focus on developing your style. Always remember the end game. What is the video for? Does it meet the needs of your audience? What do you want your audience to do after watching?

After you have those answers established, the camera will be effectively focused on your **S.T.A.G.E:**

• **S-social skills:** How you communicate and act with your audience

Despite the term audience, remember that when you are making your video, you are making it for one person: ideal client, volunteer, donor etc. People are craving authentic conversation, therefore, the better you are at relaying your message in a conversational way, the more people will be drawn in.

- **T- talent:** A result of genetics or training or both.

I do not want to belabor the difference between a talent and a gift because I don't want you to feel like you have to discover a hidden talent; instead I want you to strive at getting better at what you are already doing. If you can sing, dance, etc., and then incorporate those talents into your videos. Your audience will love it.

- **A-abilities: They can be learned.**

I have the ability to create video. I do not call my video-making skills a gift or a talent because I had to learn from the basics and train to become professional. If anyone can find any of my earlier work that has been buried in the depths of the internet, you will definitely understand what I am talking about. I use everything I have and continue to learn; I still take courses to make every video I do awesome.

- **G- God-given gifts:** Can't be taught.

"Comedy is the one profession that is nontransferable. Comedians can be great actors. Great actors can't be comedians."—Steve Harvey

If you want to stand out and get your message to spread like wildfire, you must tap into your God-given gifts. You can sing, dance, write, act, etc. but what makes you stand out from other videographers is your gift of creativity. Take all of your talents, skills, and abilities and use them to display your gifts.

• **E- excellence:** The quality of being outstanding or extremely good.

My telling you to pursue excellence in your video does not come as a surprise. But it is easier than you think. Video has come a long way. The smartphone we use daily has more powerful technology than the computer that helped send the first man to the moon. Because of this technology, your audience will be unforgiving of bad lighting and bad sound. You can get away with a few mistakes on social media because people are there to socialize and have fun but if you want to interrupt their fun for a special announcement, *it had better be good.* Your level of excellence shows the type of respect that you have for your business, mission, or movement.

It's Pre-SHOWTIME!

Time to get those *reel* muscles working. Your

assignment, should you choose to complete this mission, is to do 10 one-minute videos for 10 days. *Do not* show them to anyone. While you are completing this assignment keep in mind that you are *not* showing these videos to anyone.

Now is the time to say what you want and be who want. Don't waste this opportunity. Amp it up, be silly, be dramatic, be crazy, and be you 10 X. Be who you are when no one else is watching. This will be easy since you know the videos will not be seen; your head will allow the *reel* you to come out.

You will know when you are ready to face the camera when you are no longer zeroing in on your flaws and are focusing on the message. You have made the shift from being critical to being constructive with your critique. Only when you have made this shift, will you be able to ask the most important questions about your videos: Do you like this person? Would you like to get to know this person more? When the answer is *yes*, it is time for you to drop your message.

PART 3:

Drop Your Message on Millions

10

What Do I Say?

You did it. In the words of Steve Harvey, you made the jump into your destiny. You have done all the work required and decided to stop being a secret; now you are ready to take your message to the masses. This part of your journey can be painful when no one seems to be listening or no one seems to care.

I am sorry to tell you it will take more than a pep talk at this stage to keep you motivated because research reveals that being ignored causes the same chemical reaction in the brain as experiencing a physical injury. I find it amusing that because of this research we have yet another issue that can be solved by popping medication. But if you relate to this type of pain, don't reach for the Tylenol just yet because this section is offering a cure to *what ails ya*.

To Script or Not to Script

A well-written script is the foundation to a successful video. I realize that some people believe that they do their best when they are unrehearsed but I strongly caution against this. Are you authentic and real if you use a script? Absolutely! Do you think you are not smart because you had to study to earn an "A"? Well then don't fall for the trap of believing you are being a fake just because you prepared a script.

Monopolizing the conversation is one of the biggest mistakes entrepreneurs make when they create a video. Your audience perceives you as obnoxious, promoting stories about how great you are. With the power of a screen swipe, your audience will make you magically disappear. No matter how good or needed your services or cause, people are only interested in: *What is in it for me?* A good script therefore will convert your audience and keep them from catching a snooze.

The first rule of script writing is always: Keep It Short! People's attention span is now compared to a goldfish's—seven seconds. A script keeps you on the right track. Unless your video is an opt-in or a video lesson, keep your video to less than two minutes (350 words is approximately 2 minutes.) If you find your message/script is way too long, don't panic. Use your long-winded talents and break up your message into a series.

Although you have prepared a script that is no reason to sound scripted. You are going to have to practice before you press record. No, that does not mean you have to spend your free time trying to memorize your written masterpiece. What that does mean is that you will have to develop the art of being conversational. The following outline of sales script will give you an idea of how easy it actually is.

Paragraph 1- Introduction (What the video is about.)

Paragraph 2 –State the obstacle your audience might be facing (Why they should listen to you.)

Paragraph 3 – Talk about your 1st point

Paragraph 4 - Talk about your 2nd and 3rd points

Paragraph 5 - Summarize and Call to Action

This brief example does not look very conversational, I know. So when it is time for you to write, I want you think of one thing: a friendly phone call.

Me: Hey, it's me.

Her: It's so good to hear from you. I was hoping you'd call.

Me: Just dropping you a line because I noticed you signed up to chair the new committee. I thought you could use some help.

Her: Omgosh, that would be awesome, yes I would love your help.

Me: Well I have been doing this type of work for a few years so I think the first thing you should do is _____ and always remember to _____and if you don't remember anything else, always _____.

Her: This is so going to help me so much, I am so glad you called.

Me: No problem, I am sure that if you follow these tips it will help you get started and even make things easier along the way. If you want more help, call me, text me, or you can join my club.

See how easy that is, you have been unknowingly writing scripts for years. Here is how you would write a script for a video using this same outline.

Paragraph 1- Hey Shonda Carter here, your storyteller with a camera; today we are going to learn how to create the perfect script.

(They haven't clicked off--translation: This is just what I needed, I am so glad to hear from you.)

Paragraph 2- If you are here that means you are ready to dive into this whole video thing. But I know sometimes we can get stuck on what we are supposed to say. That was one of the problems I faced when I started doing videos 15 years ago.

(They are still listening--translation: You read my mind, this is just what I needed, I am so glad you are here.)

Paragraph 3- One of the first things you should do is introduce yourself and let your audience know what you will be talking about.

Paragraph 4- Second, you should establish your expertise so that the audience will have confidence in the information you are sharing. And finally it is always a good idea to share at least three great tips and a strong call to action.

Paragraph 5- Well that just about sums it up, these are the same tips I used when I started making videos so I know they are effective. If you want to learn how to transform average video into show stopping productions stop by my website at www.ShondaCarter.com and schedule your free video marketing strategy today. I only have a few spaces available so don't wait too long. Remember it's hard to be a success if you're a secret! See you next time.

(Stayed to the end-Translation: I really like her. She is great. If she can give this type helpful information in two minutes, I can't wait to see what I can learn talking with her on the phone.)

You see how easy that is? I am not naive enough to try to make "one size fits all" but I have learned that although it makes things a lot easier with a script, not everyone can be scripted. If you are one of those people, then a great director is a necessity. A great director is well versed in the art of patience, practice, and perseverance; therefore she can help you make a dynamic video in a way that is natural to your style. You will still need a script; the format will be different but the results will be the same.

If you still can't see the necessity of a script, look at it from a different perspective. Don't call it a script, call it a plan. I am stressing the importance of this because no one wants to tune in to you for help and hear you rambling on and on. Even if you manage to stay under the 2-minute mark, if you don't grab that person and deliver the information she wants, your 2-minute video will be played for 20 seconds--and that is if the person is nice.

So now that you know what to say, the next question is: What should your videos be about? You will notice that the above examples were not used for promotion, but helping. There is good reason for this; these are the types of videos you will be doing the most. Remember the words of Theodore Roosevelt, *"No one cares how much you know, until they know how much you care."* Your audience will know that you care if the majority of your videos focus on their hurdles or issues but not if you only tell them how great you are. Once a basic line of trust is established, they will want to get to know you better. This is when your audience will specifically begin to search for your videos. Here are a few examples of the types of videos you can create so your potential customers can get to know you better:

• How did your company or organization develop?

• What motivates you to keep going?

• Why do your customers find value in your work (client testimonials)

• Behind the scenes peek of your company/quirky

employees.

• A great concept you embrace (What do you stand for?)

• And yes (when the time is right) Promote!

11

Sharing is Caring

"People don't care how much you know until they know how much you care"—Theodore Roosevelt

My first experience with a business coach was a bit of a disaster as I had no idea what a business coach/mentor was or how I was supposed to interact with one. All I knew was that I needed help to get started in my career. I thought my prayers had been answered when I clicked on a video of beautiful, successful woman who seemed to be speaking directly to me. She told me that I could create a successful business and live the life of my dreams. I was *asking* for help and there she was. I followed her (social media--not stalker-ish) everywhere. And I eventually signed up to work with her. Well, I wish I could say that when we parted ways, I was a six-figure thousandaire but that wouldn't be true. I was feeling a bit on the bitter side. I guess I played the blame game for too long because I started to hear that still small voice offering a commentary on my *woe is me* behavior.

Despite the lack of tangible results then, I have discovered a new appreciation now for my first coach. She at least cared enough to help me find the *reel* me. A few years back, the term *coach* was not as popular as it is today. Despite my need for help, I had no earthly idea of where to look or if the help I needed even existed. That video that I watched years ago was much more than the spark that started me on my journey, it also represented time, effort, and energy spent trying to reach me.

I know that people believe that she was just advertising as usual but it really doesn't matter what you think because I was the one she was trying to reach. That means, mine was the only opinion that mattered and to me; that five-minute video was not an ad, it was the answer to my prayers.

I listed the stats so I am sure that you know by now video is the fastest way to get your message to the masses. But that is only if it is watched and shared. To accomplish this, you must create messages that not only reveal your passion for your business but your customers as well. Today's consumer wants to be invited backstage to socialize. So are you ready to create a powerful video marketing strategy that will create Super Fans that spread your message worldwide? If your answer is yes, let's start where all great relationships start--emotion!

People don't do business with companies; they do business with other people. The fastest and easiest way to humanize your business is to develop a video marketing strategy dripping with emotions. This does not mean you have to create productions that cause runny mascara, tear-soaked tissues, or a lump in your throat. There are plenty of other emotions besides *sad*, and the following list is guaranteed to evoke a response from your audience:

Exhilaration	tement	Love
Amazement	Warmth	Humor
Inspiration	Nostalgia	Surprise
Contempt	Uncertainty	Shock
Sympathy	Frustration	Serenity
Fear	Desire	Joy
Delight	Admiration	Optimism
PassionExci	Anger	Validation

And if you want to create and strengthen brand loyalty while getting your message shared by millions, let's take those emotions to where the magic happens—social media.

It's amazing how long it took me to learn some of the science behind social media sharing; I thought people shared because they liked to do so. Imagine my surprise when I discovered that there is a set of chemicals behind the scene dictating whether your video will go viral, or die in obscurity. These chemicals are known as dopamine and oxytocin. While I have heard these chemicals mentioned on The Dr. Oz Show, I had no idea just how powerful they are. According to PsychologicalScience.com, dopamine causes social media, and checking email can become more difficult to resist than cigarettes and alcohol. Scientists believe the rise of oxytocin during the use of social media is equal to the hormonal spike some might feel on their wedding day. But what does this mean when it comes to sharing? If the activity is driven by the reward center of our brains, if the stimuli are right, we can't help ourselves, we will have to share.

I hate to ruin the word *sharing* because I know it conjures up a wonderful unselfish image but 68% of people say they share not because it will benefit humanity but because it shows people who they are and what they care about (www.nytmarketing.whsites.net). Our *shares* don't only explain or represent us, they make us look good. In the world of social media, when we share cool and funny materials, it makes us look cool and funny.

So what do all this research and science and facts and figures mean when it comes to video? Only everything. Video is beyond hot right now thanks to social media giving preference to it and people being psychologically driven by it. And while The Ice Bucket Challenge was for a good cause, it also proved to be one of the most powerful aspects of social video: Messages are so easily shareable, they can spread rapidly.

You can't just throw anything out there and expect it to be shared just because it's a video. No, there are regulations that even though unknown must be followed. So let's find out what they are.

12

Right place...Wrong time?

There is a story I heard from a well-known minister in Christian circles and TV, who said that the Lord told him to plan a conference that would concentrate on the subjects of faith and prayer. Being ever obedient, this minister began the process; he picked a date, booked a venue, and passionately studied so he could be prepared for what he believed would be *an awesome move of God*. The day of the event, the minister was heartbroken to see that no one outside of his small congregation came to the event. During his prayer time the next day, the minister made it clear that he was a little upset at the dismal turn out. He said the Lord told him that happens when you fail to tell other people about the event. The minister was shocked and said, "But Lord, you told me to put this all together, I thought *you* would tell everyone." The Lord replied, "If people could hear me loud and clear and did everything I told them, why would I need you?" Moral of the story: Even the Lord has to use public relations to get His message out.

During the Question & Answer portions of my presentations, one question never fails to arise: Where should I put my videos: on Facebook or Youtube?" I always take the time to explain the benefits of both. Because social media is bursting with competing forces (Facebook and Youtube are not *besties*); if you want to get your video watched on Facebook, then always upload it straight to Facebook. Facebook does not necessarily interfere with content shared from other platforms but it does not make outside videos look like something that should be watched. Facebook averages more than four billion video views per day but the display or thumbnail for a Facebook video is six times larger than the thumbnail provided when content is shared from other video sites (including your own website). Facebook videos look bigger and better and scream: Watch me! All others get lost in the timeline shuffle, if seen at all.

Because Youtube is owned by Google, your videos will usually appear on the first page when conducting a Google search. According to SimilarWeb.com, Youtube is the second largest search engine and the third most visited site on the web.

Right place, wrong time or sometimes, wrong place, right time. I remember that saying when I when I try to figure out when to share my videos for maximum impact and share-ablity. I did my own research called *hit or miss* and was grateful to discover that RadiumOne did a more reliable study of the best and worst times to post online. According to their research, the two peak sharing times during the day are:

- Between 10:00 a.m. and noon and then between 8:00 p.m. and 10:00 p.m.
- The worst time to share, if your goal is to maximize clicks, is between 9:00 a.m. and 11:00 a.m.
- Sharing and clicks remain consistent from 1:00 p.m. and 6:00 p.m.
- The hours between noon and 2:00 pm are at maximum for the most shares with the highest clickthrough rates.

Best Times for Sharing on Social Channels:

- Twitter: 1:00 p.m.
- Facebook: 5:00 p.m.
- Pinterest: 11:00 p.m.
- Google+: 10:00 a.m.
- Email: 7:00 a.m.

Producing creative content and scheduling at the right times can be time consuming but the results will be worth it, especially since Instagram has 100 million active users, Twitter 135+ million, and Youtube over 1 billion active users each month. Even if you get only a fraction of those visitors interested in your message, you will amplify your reach over 1,000 times.

Is This Thing On?

What happens, though, when you follow a perfect schedule or a perfect video format and you still are not seeing the results you want? My advice is to keep going. Please, do not stop. You never know who is watching. I often am recognized at network events because of my videos. But early on in my career, a guy approached me and began talking like we are old friends. I did not think too much of it, until he told me how much he *loved* my videos. I tried to hide my puzzled look because back then, the only *likes* and *shares* that I got were exclusively from friends and family. Eventually, I realized the legends were true and I was speaking with a *lurker*. Even though this term may bring up feelings of creepiness, it is not. The term *lurker*--when you are using it in reference to online activity--means someone who is interested in and observes your videos but does not let you know that they are there (no *likes* or *comments*).

I know that if you are doing business online,

lurkers can cause a sense of frustration, especially when you realize the shocking ratio reality. According to the Neilson Norman Group, 90% of your followers are lurkers, another 9% occasionally participate, and only 1% regularly engages. While the percentages may be shocking, the reasons why are typical. Some of the same excuses given for avoiding a camera are the same for lurking. Some fear being mocked. I experienced this fear when I bravely began participating in online groups. I started out humbly by thanking everyone in the group for their participation and telling the members how much I had learned from them. I laugh about it now but I made the *mistake* of thanking God for the group. Well, by one member's reaction, you would have thought that I had burned her house down. She quickly cursed me out and warned me that "there is no God and even if there were, he would not give a (bleep, bleep) about your stupid business." I am happy to say that some of the members were just as confused and startled by the response as I was and immediately started to defend me. She was reported to the admin and swiftly banned from the group. The sad part is, after all of the drama that happened, I never posted in that group again.

While I have no problem participating online now, I often wonder what kept me in lurker status for so long. Intimidation was definitely a factor. I never fully embraced the title of *expert* then so I did not think that it was a good idea to give advice concerning the questions that were being asked in the group. And I didn't want to ask questions because I didn't want to look like an idiot, so staying silent and soaking it all in seemed like best option at the time.

If some of your lurkers feel like I did, they may also be worried that they will be misunderstood. This is the danger of catering to only 1% of your audience. After all, they are driving the momentum of your brand. Or are they? It would be a mistake to believe that just because they are quiet, they are to be ignored. As marketers, we need to understand that the listeners are just as valuable as the talkers, because eventually, they may feel safe enough to participate and ultimately turn into a *sharer*. If we take the time to understand the stories and conditions of the people we hope to convince, we have a better chance of creating a story they want to hear. As marketers, we need to understand what could happen if we invest as much time understanding our audience as we do in being heard.

13

Going Viral: Chasing the White Whale

Most entrepreneurs want their videos to go viral. And most videographers wish that we knew the formula to make that happen. Yes, despite the numerous blogs, how-to videos, and lesson modules that promise the dream scenario of going viral, there is no recipe for creating this type of attention. Even spending money for advertising is still no guarantee.

According to TechnoPedia, the term viral video is a clip of animation or film that spread rapidly through online sharing. Viral videos can receive millions of views as they are shared on social media sites, reposted to blogs, and sent in emails, etc. There are so many failed attempts at trying to recreate viral type success that I spend the majority of my time trying to stop a client's attempt. If your goal is to get more clients, income, and influence, a viral video will not get you close to those goals. It will almost have an opposite effect.

Viral Video is instantly famous because it gives you a quick laugh, shock, or major dose of inspiration. It is not designed to make sure that you or your brand is remembered. Anyone know the real name of the Chewbacca Mom?

Seeking to build an entire marketing strategy around a viral video is the equivalent to starting a company with the purchase of lottery tickets. So let's stop chasing the 15 minutes of fame and get down to the work of supporting and sharing your message with strategic video. While I can't guarantee that your video will go viral, the following information will get you as close to it as possible while creating the impact you desire for your message.

Make It Worth It

Before you try to share your video with the world, make sure it is something you would share with a friend. When I call one of my friends to tell a story, I know how they will react. Whether they laugh, cry, or get angry, I already know this in advance because I have built a relationship. I know my information will evoke some type of emotion. The conversation that we have will benefit both of us. If you insist on making videos that focus exclusively on what you think is important instead of invoking emotions, you will probably end up hearing the echo of a video dial tone.

Understand the Trends

"Stories sell, tell yours well" is the latest quote marketers are using to emphasis the importance of storytelling. Although storytelling has been around since the beginning of time, it is being hailed as the cure for everything concerning social media. Video is no exception. Video is shared not only because of its ability to hit the heart but also its ability to give us the power to show, not only tell. Don't waste your time talking about products or results—let the pictures tell the story and the subconscious will do the rest.

Subtlety is the Key

People usually don't share ads (unless they are already trusted), which is why it is never a good idea to make your logo the star of the video. Because you want to be remembered without being overly promotional, showing your brand's personality throughout the video is essential. How do you do that? Once again, ditch the cheesy sell for the subtlety of entertainment.

14

Get Rid of the Box

"Once people make your story their story, you have tapped into the powerful force of faith."
—Annette Simmons

One of the best ways to get your message shared is to stop trying! Stop putting your brand in a corner by doing things like everyone else. Stop trying to market and stimulate buzz. It is time to ignite passions and start a movement. Creating a brand that stands out from something other than money or stock options is the fastest way to disseminate your message. Your audience won't be difficult to find if you are ready to move, inspire, and compel them with a brand that is working hard to change the world. When you focus on spreading your message instead, the focus shifts to mobilizing people to get behind your shared message. Even if you know that your brand will not change the world, if you want to start a movement, you have to have believe in what you are doing. The stronger your beliefs, the stronger your brand and the stronger your stand and that is because people want to belong to something that is bigger than them. A movement is the reason people line up around the block to buy a new product. No, the product may not instantly heal the world but the message behind the brand means something to its loyal followers. It breaks my heart to see people spend time and money to create videos that no one will ever see. Because of social media, everyone has their own platform, their own audience, and their own channels; if you want them to share your message, there has to be a reason. The reason must be bigger than *you* and bigger than your audience. If you can show them something bigger, something inspiring, something worth caring about for more than two minutes, I guarantee that you will start people sharing and they will start listening as well.

Showing—rather than explaining—your movement will keep your audience more interested. Video is the perfect way to educate your audience on the importance of your cause, and what is in it for them if they join the movement. In under three minutes, your audience can experience an intimate understanding of who you are trying to help and how much of an impact can be made when you and your audience work together. The use of video for this type of marketing is not only limited to explanation. It can also be used for celebration. There is no better way to celebrate and thank your supporters than with a unique *thank you* via video. Even if your product or service is dull, the power of video and story behind why you sell them, will make your goods instantly compelling.

Part 4:
REEL IN THE DOUGH

15

Can't Afford Not To

People today get most of their information online. As our appetite for the Internet grows, video is also ready to grow as businesses desire more of an online presence. Despite the information available, listing all the benefits and reasons that video marketing is essential, people still decide to decline video services. The reason usually given: "I just can't afford it." Usually I respect a client's decision and leave the meeting with a smile but internally I am screaming. I am internally screaming the words I want to say: You can't afford not to! Unfortunately I have realized that some clients don't recognize the value of great video until after they find someone cheap, after they waste dozens of valuable hours trying to take a stab at their own video editing, or after they are still the best kept secret in their industry.

While this book will help you with the basics of video and editing, there are a few problems you will not overcome without hiring a professional. As a professional videographer, I strongly believe in

Abraham Lincoln's message: "He who represents himself has a fool for a client."

Two scenarios represent the need for a professional videographer. You agree to teach a class or make a presentation. The first mistake that you make is considering it to be routine—no big deal. By the end of the presentation, you realize you *rocked it* and your audience's raving comments confirm your insight. You will *kick yourself* because you didn't have your presentation recorded because you will realize the endless amount of clips that you could have used for Youtube, Twitter, Facebook, LinkedIn, etc., and the website embeds guaranteed to increase your Search Engine Optimization (SEO).

Ah, but then there were those times when you asked a friend to record your event. All you need is someone to aim the camera and press a button, right? Wrong! You will get footage of the event and you will spend many hours trying to *save* that footage to make it usable.

I am not going to tell you that you are wrong if you truly believe that video is not worth the price. But once again, I am going to ask that you look at it from another perspective.

Video producers are not just recording for a few hours on a designated day. They also have to edit the hours of footage (which can take up to 60 hours) to create video that captures the essence of your brand; ease your fears; answer all your questions, phone calls and emails; and continue to run a business while being at your beck and call. So many people are under the misconception that they will be paying hundreds or thousands for *one day of work*. You have to realize that there is work that must be done before and after the big shoot.

Also consider that your visual image/branding is the first impression that you make when people visit you online. Professional video equipment is expensive as is running a business. If your videographer is charging a *too good to be true* rate he or she is cutting costs elsewhere. It would be a shame to find out that one of the areas skimped on was the production quality of your videos. Be careful if a low price is the only thing you are looking for to determine who you hire.

If having an online presence is not crucial to your business then video is not worth your investment. If you don't need to increase traffic to your website or landing page then video is not worth your investment. If you don't need sales, skip this craziness all together. You are one of those people who has proven your point; video marketing is *not* for you.

If you can't afford to hire a professional at this time in your business that is completely understandable (been there) but don't ever believe that video marketing is not worth your investment. No matter if you hire out, or have to do it yourself, it is always going to cost you money. I now ask you to invest some time as we learn the basics of all things technical.

16

Don't Get Left in the Digital Dust

A successful video marketing campaign requires at the very least five videos per month. I am not talking about major studio productions. The kind of videos you should be doing can be done at home or on the go. I use these types of videos as an opportunity to share my knowledge about brand stories plus it shows off my wacky personality. Remember, the opt-in website video, strategy sessions, major event promos, and sizzle reels should be shot and edited by a professional. But even if you don't have a professional on speed dial that is still no excuse to put your video marketing on hold. Speaking of excuses, let's cut through the clutter so we can keep you from being left in the digital dust. This list is not too long since we have already tackled a few of these areas when we agreed to stop being a secret.

1. **I don't have a good camera**: I hate to admit this, but an iPhone is better than the

camera I spent a thousand dollars on a few years ago. Yes, high-quality, HD video is wonderful to view and I have my eye on a few 4K cameras myself but for capturing exclusive behind the scene footage or anything spontaneous or funny, your audience won't care what camera it was shot on, only that they are able to be a part of the experience.

2. **I don't have a professional set:** Not really necessary. I have access to a professional studio and I barely find the time to use it. Fantastic lighting and good sound are the only accessories you need. If you feel that you must have some sort of set dressing, get a fabulous chair and a couple of non-distracting props and shoot away.

3. **I don't know how to edit:** A good editor is priceless but we all have to start somewhere. I started with absolutely no training and the software MovieMaker on my husband's PC. Now my 15-year-old can edit a school project with iMovie and make it look like a *Star Wars*™ film. If you don't have an iMac or an iPhone, do not panic. There are wonderful smart phones and plenty of free editing apps to get the job done. I must caution you, however, about choosing editing software based exclusively on price. If you are planning to do more

than piecing together basic cuts, I would advise you to do the research and make the investment required to get you started making video that will reel in the dough.

I understand that unless you are a professional or highly interested in film editing as a hobby, the thought of sitting at your computer trying to edit video together is not exactly an ideal situation. This is why I strongly caution everyone to take care of as many problems as possible before the camera starts rolling. Bad sound and bad lighting not only *scream* that you are an amateur; they are distractions that no audience forgives.

Nothing beats natural light, so filming with a window *in front of you* (never behind) will always offer you the best solution. A lavalier microphone will help enrich your sound and you won't have to break the bank. You can find one on Amazon® ranging in price from $6.50 to $600.00. On Amazon you can also find a mini phone tripod to keep your camera movements to a minimum.

There are few common mistakes I must warn you about when creating *quick* video. Most people believe that it is easier to talk for a few minutes looking straight at the camera to avoid having to edit. While this does seem like an easier way to get your video done, it will end up really boring.

Using the same straight-on, death stare camera angle shows a lack of imagination. The audience scrolls right by this type of video, unless the star of the video already has an established audience. Because you produced your video *quick and easy* means nothing if it has no audience. To mix things up with your camera angles use the five-paragraph script template discussed in Chapter 10 as a guide.

Paragraph 1- Use a straight-ahead camera angle for your quick introduction and get straight to the point of your message.

Paragraph 2 – Talk about your first point, but *show* b-roll video (supplemental or alternative footage) or pictures to illustrate your message. *"Look it's a bird, it's a plane!"*

Paragraph 3 – Use a straight-ahead camera angle and talk about your second point.

Paragraph 4 - Place your camera to the right of you but keep looking straight ahead as you speak and talk about your first point. I know it will feel weird but trust me it will look great. If you are uncomfortable with  this angle, another option would be to film from a different location than your previous shots.

Paragraph 5 - Use a straight-ahead camera angle to summarize; mix it up by using graphics to display your call to action.

See how easy that is? Now you have five shots to put together that will keep the action flowing and force the viewer to readjust to the action, thus, keeping him or her engaged. Best of all, you won't be overwhelmed trying to remember an entire script because each shot is just a couple of sentences.

You've filmed your video; let's edit. Things to remember:

1. Don't use fancy transitions in between shots. They might be fun for you to do but they are distracting and scream that you are an amateur.
2. Use music. Music keeps the action flowing and sets the mood for the video in advance.
3. Don't use music that *wraps up* as this could signal to a viewer that the video is ending and cause him or her to exit prematurely.
4. Do not spend precious time reviewing the points you covered or giving the audience time to write down notes. If your viewer missed anything they can pause the video-- or even better--watch the video again.

The editing process is where the magic happens but keep in mind there is no such thing as wasted video. Turn your mistakes into a teachable moment or a chance to have some fun. Share your bloopers. Bloopers are entertaining and show authenticity. Nothing connects you to your audience more than showing your frustrations and challenges as you teach what your audience may encounter or already encounters.

17

Effective Video Marketing Strategy

Hannibal Smith is not the only one who loves it "...when a plan comes together." If you want to make all this video work worthwhile a plan is exactly what you need. Even if you are just beginning, please don't make the same mistakes that I did by pouring your heart, soul, and time into a video only to find yourself heartbroken because it just sat there alone, rejected, unwatched, unappreciated, and unloved—you get the idea. While your campaign strategy may conjure up incredibly detailed research and details, you can relax because, if you have done the work in the previous chapters, the majority of your effort is done.

People frequently ask me to tell them the return on investment (ROI) for video marketing. While I understand that it makes some people feel better when you can provide percentages and specific numbers for their investments, I am not able to give those types of numbers during a presentation because you are the one who determines what the ROI will be. For example, if you want to fill your event, then a full room is the return you are seeking. To determine, what your ROI will be you must map out a purposeful plan:

- **Who and What are the videos for?**- Are you trying to educate, entertain, explain how to use the product, etc. Remember your brand's vibe and your audience's need (not what is easy or what you like best) should be the determining factor.
- **WIFT (What's in it for them?)**- Why should your audience invest the time to watch your video? What will they get in return?

Once you answer these two questions, you will be able to create your video topics and types of videos for your website--not social media. While social media is great for getting your message out, their ultimate goal is the same as yours--keeping people on their site.

- **Opt In Video**- The homepage of your website is the perfect spot for this 2-3-minute-long video that captures your visitor's name and email address and permission to contact them. Offer a freebie (free report, e-book, etc) as a reward.

- **About Me Video**- The About Me page is one of the most useful and popular pages on your website. Don't waste this page with a résumé. Use a video to introduce yourself to your audience, inform them of the value they will get from you, why your site/business is unique, and a behind-the-scene look. About Me is yet another opportunity to extend a call to action and extend an invitation to work with you.
- **How-to Videos**- (Services page) Text can be tricky and your clients will appreciate your explanation about how to use your product or how to start the process of working with you. People usually gain more understanding when they see or hear an explanation. Another bonus: you can also explain why you are better than the competition.
- **Testimonials**- (Our Work /Testimonial/ or every page on your website) "Nothing draws a crowd quite like a crowd"—P.T. Barnum. Adding video testimonial is proof of your brilliance and the best way to generate more selling power without sounding sales-like. According to marketing researchers, up to 70% of American shoppers say they look at the product reviews before making their decision to buy online.

What about all those wonderful showoff, show up, and showcase videos you just learned how to do? Where do they go? The question is: Where don't they go? Here are a few suggestions:

- **Website:** Because you use your website for business, doesn't mean it has to be all business. Don't use your wonderful personality exclusively on social media.
- **YouTube:** You need to get seen, get heard, and get known so you can get paid. YouTube will help establish you as an expert and get your fantastic videos found.
- **Email:** Use video content in your email campaigns to increase Clickthroughs.
- **Presentations**: Incorporate video into your speech to keep your audience engaged. Show, don't tell.
- **Social Media:** Don't want to be repetitious but reread chapters 10-14 and use a strong call to action to bring fans to your website.
- **Print Advertising**: Yes, a Quick Response (QR) code linked to your video can be placed on your printed flyer or business card to give you the best of both systems.

Now is the Time for ROI

A strategy that tracks your growth will advance your brand--a strategy to improve your marketing and keep you looking forward. Because video is distributed via a *player* you can easily measure performance through a video marketing platform.

Video Analytics- Look for play count and drop off rates. If 90% of your audience does not make it to the end of your video that means they are missing your call to action and, therefore, you are not getting a return.

Click through Rates- Determined Clickthrough rates by doing a split test with your email. Measure how many open when the subject line contains the word video or not. This can also be done with website pages with video versus those without.

Measure your CTA's– This may be the most solid way to measure your ROI. You can attach a Google code to track the number of people who watch your video and compare it to how many actually sign up. This will determine your conversion rate. Facebook ads also supply this information.

Despite how hard you try, there are some things you are unable to measure with facts and figures such as the excitement or emotional response to your video. You must have a plan, then, if you are adding video to your marketing strategy to make money. Going viral is not a strategy. Neither is expecting one video to solve all your needs. No matter what kind of video you produce, your video must always start with an objective, as there are just so many goals a video can achieve.

Wave of the Future

When you visit a social media site, you will notice that video no longer has the title of futuristic and cutting edge. It has moved from the status of innovation to the status of being expected. Because people are able to create instant fame and build entire careers around their passions through social media, our once peaceful timeline looks like Time Square with its flashing array of billboards. This has driven some entrepreneurs to create endless barrages of live streams and *look at me* videos. While the latest trends could be used as effective tools, I ask that you try not to be everywhere doing the latest version of everything. Instead concentrate on projecting ahead of the trends and focus on what drives actual engagement by making quality a priority, not quantity. The long-term benefits will include a respected reputation, brand loyalty, and more money-making opportunities. To do this, you must adapt to the newest advances. Video is no longer the future; it is now. It is not a trend. Some experts even predict that it could replace text completely. This doesn't sound plausible although I never thought I would hear teachers telling their students not to write in cursive (bloomberg.com). We may not use video exclusively in the future but clearly video has become an essential part of our lives.

Appendix 1
So What Happens Now?

One of my favorite memories is sitting with my father watching *Star Trek*™. While I always related to Mr. Spock, Dr. McCoy came to mind when I was advised to include a "how to work with me section" in this book. I believe my first thought was "Darn it Fred (Jones), I'm a videographer not a car salesman!" Thank goodness that line of thinking was interrupted by yet another memory, "Where no counsel is, the people fall: but in the multitude of counsellors there is safety," (Proverbs 11:14 KJV). Otherwise, I would have failed to remember that learning to make great video doesn't happen overnight.

You can visit me online at www.ShondaCarter.com for more tools and resources designed to create shostopping video. But, as a special treat I created something specifically for you so we can continue this journey together--ShoStopper Academy.

Whether you are called to a specific market, a movement, or a ministry, this is a month-to-month course will expound on the power of video to help fulfill your mission. We will cover: visual branding, script writing, avoiding tech terror, developing a video strategy, creating speaker reels and any other conundrum concerning the skill of video marketing--including the accountability needed to get those videos done.

I want to help increase your skills so you can make compelling videos, brimming with creativity, that convert like crazy. When you're ready to "link up", go to www.ShoStopperAcademy.com.

I have shared so many quirky thoughts and vulnerable moments, now I would love to get to know you as well.

SHONDA CARTER
Creative Captivating Comedic

Appendix 2
The Videographer's Legal Checklist

Videographers must understand the legal principles about copyright, privacy, and permission before venturing into producing videos. The checklist below is a general overview of the most pressing legal issues. Creating and sharing videos has never been easier, more fun, and more profitable. Know your rights.

1. **Is copyright automatic?**

The moment the ink dries or the moment the flash burns out, the captured content has been copyrighted. Only tangible forms of expression (e.g., a book, play, drawing, film, or photo, etc.) are copyrightable. Once you express your idea in a fixed form—as a digital painting, recorded song, or even words scribbled on a napkin—it is automatically copyrighted if it is an original work of authorship. No official step is required for this minimal level of protection. It is important, however, to take the next step and record your content with the Library of Congress.

2. Did you obtain a written photo release?

The safest way to operate is to get a written release from everyone when they come into the room. This is critical because obtaining permission upfront will avoid those who may want to hold off on giving permission once they see the images of the event or the final outcome (e.g., a book cover, CD cover, drawing, photo, film, etc.).

3. Do you provide a copyright notice?

Copyright is a form of protection provided by U.S. law to authors of "original works of authorship." When a work is published under the authority of the copyright owner, a notice of copyright may be placed on all publicly distributed copies or phone records. The use of the notice is the responsibility of the copyright owner and does not require permission from, or registration with the Copyright Office. Use of the notice informs the public that a work is protected by copyright, identifies the copyright owner, and shows the year of first publication. Furthermore, in the event that a work is infringed, if the work carries a proper notice, the court will not give any weight to a defendant's use of an innocent infringement-Copyright Notice defense—that is, to a claim whereby the defendant claims not to have realized that the work was protected. An innocent infringement defense can result in a reduction in damages that the copyright owner would otherwise receive.

4. **What is basic copyright?**
Basic copyright is the right to that which is copied. Copyright comes into play the moment that which is being copied, videoed, or written is placed in a tangible form. It is always good to give notice on any video, paper, or photos you want to protect.

5. **What is Fair Use**?
Fair use is the doctrine that brief excerpts of copyright material may, under certain circumstances, is quoted verbatim for purposes such as criticism, news reporting, teaching, and research, without the need for permission from or payment to the copyright holder. Many are afraid of using the works of others but there are some limited circumstances whereby it is legal to use copyright protected content. Fair use is a defense to a copyright claim. The law grants limited protections for first amendment expressions.

6. **Do you have the rights to record here? Public vs. Private?**

It is always good to get permission when recording in a private location. You will need permission from all who are in the room. It's best to get this permission up front.

7. Did you take or create your own still images?

There are many image sites online. Many of them are free sites and many are not. Often small business persons go online and grab an image and use it. Know that images are copyrighted whether or not the copyright notice is given. Once you are using someone's image and that person recognizes his or her image, she or he might contact you to stop using it. Best case scenario: Take your own still shots with your own camera.

8. Have you registered your final work?

Anyone can register basic claims to copyright, even those who intend to submit a hard copy(ies) of the work(s) being registered. Basic claims include literary works, visual arts works, performing arts works, sound recordings, motion pictures, single serial issues, groups of serial issues and groups of newspaper/newsletter issues. At this time, the following types of registration are not available: renewals, corrections, mask works, vessel hulls, groups of database updates, and groups of contributions to periodicals. For information about registering these types, see the Copyright Office website at https://www.copyright.gov

REF http://www.copyright.gov

This content is for coaching, teaching, and educational purposes only and is not a substitute for legal counsel. Should you have specific legal questions please seek and attorney for assistance regarding you specific legal need.

Authors Biography

Visual Virtuosa & Showstopping storyteller Shonda Carter is a #1 Amazon best selling author, speaker, writer, producer, director and for entrepreneurs, social innovators, and ministries looking to escape the status quo and maximize their impact.

Just a few years ago ShoStopper Productions was created by daringly quitting her behind-the- scenes career in the media industry. She designed a website, hired a coach and waited for her ideal client to arrive. She waited and waited, wasting time and making mistakes that did not lead to a successful business. When she decided to focus on creating connections instead of commercials via video—encouraging others to do the same— amazing things happened. Her client base doubled because she had the courage to be herself. She shifted from hiding to center stage. She used her knowledge, experience and brilliance from her 15 years in the business to work helping others put their genius on camera.

This introvert turned scene-stealer has co-produced and consulted on three full-length films. Her thought provoking and creative work has been screened in several open genre independent film competitions. But she is very proud of the fact that through ShoStopper Productions LLC, she creates content that egages an audience using cutting edge techniques and neither she nor her clients ever have to compromise who they are.

www.ingramcontent.com/pod-product-compliance
Lightning Source LLC
Chambersburg PA
CBHW070324190526
45169CB00005B/1727